Grades K–2

Science Activity Books

SPACE SCIENCE

Created by The Good and the Beautiful Team

Cover Illustrated by Sandra Eide
Pages Illustrated by Amanda Gulliver
Design by Phillip Colhouer

© 2024 The Good and the Beautiful, LLC
goodandbeautiful.com

Lesson 1

Cross out the items in the key as you find them in the picture. Color the picture.

KEY

Trace each line from a telescope to the object it discovered. Color the pictures.

Lesson 2 — In each row, color the two planets that are the same size.

Draw tails on the ends of the comets. An example comet is below, to the right. Color the picture to make it look like outer space.

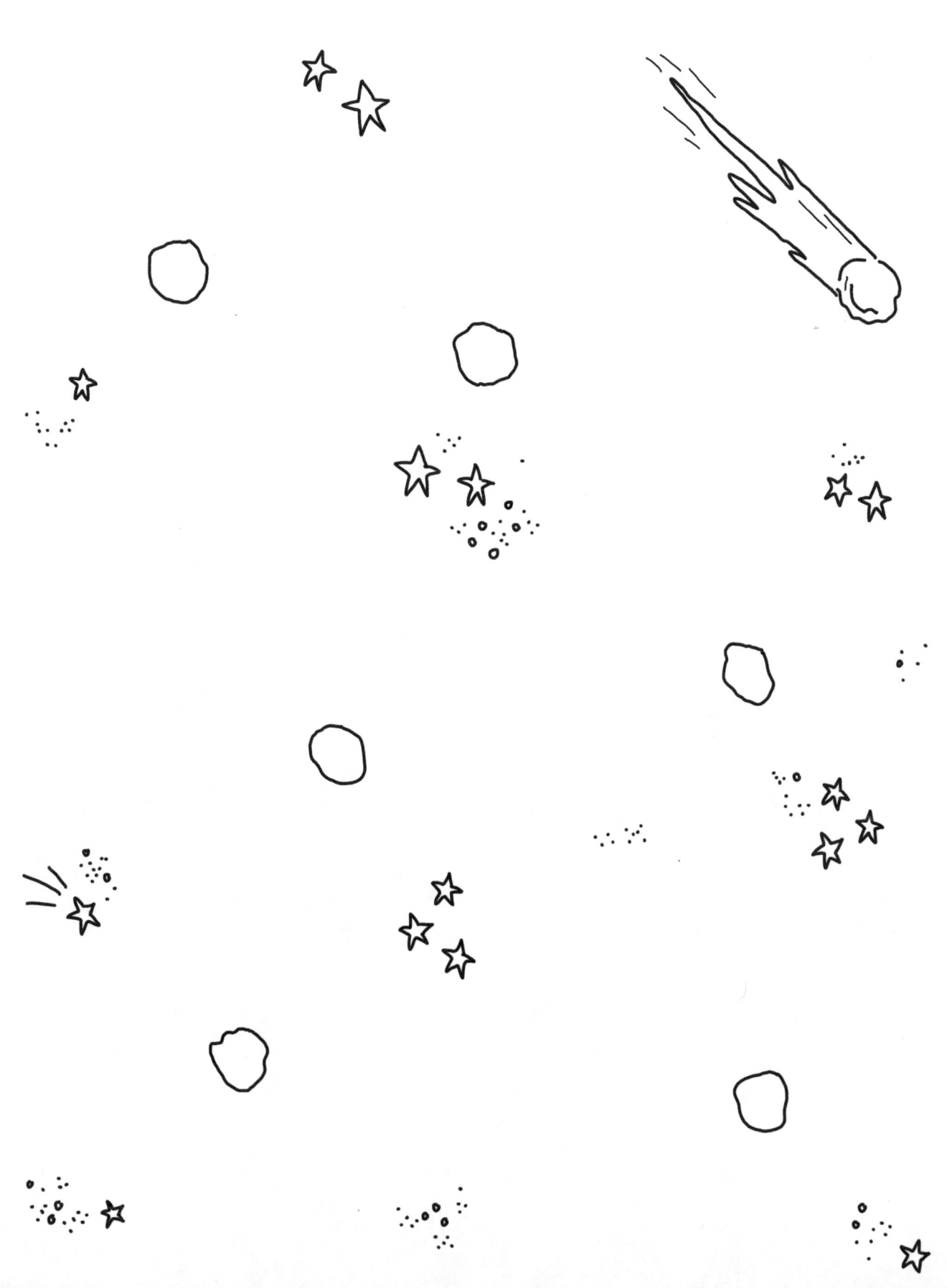

Lesson 3

Trace the lines to create constellations. Color the picture.

Color the constellations at the bottom of the page and cut them out. Tape or paste the constellations into the picture.

Page intentionally left blank

Lesson 4

Cut out the puzzle pieces. Arrange the puzzle correctly. Color the pieces. Tape or paste the puzzle pieces together on the top half of this page.

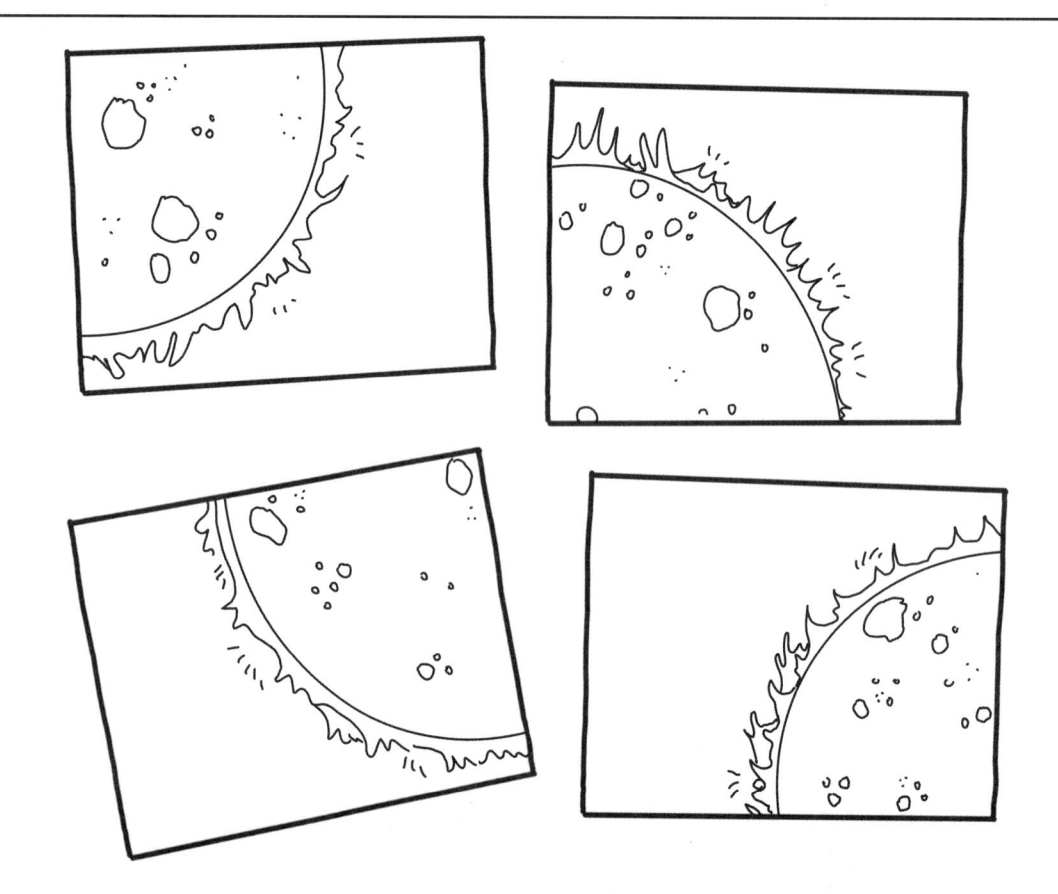

© GOOD AND BEAUTIFUL

Complete the maze from the START to the FINISH, drawing a line through each planet in our solar system. Color the planets.

FINISH

START

Lesson 5

Trace the lines to connect the matching planets. Color the planets.

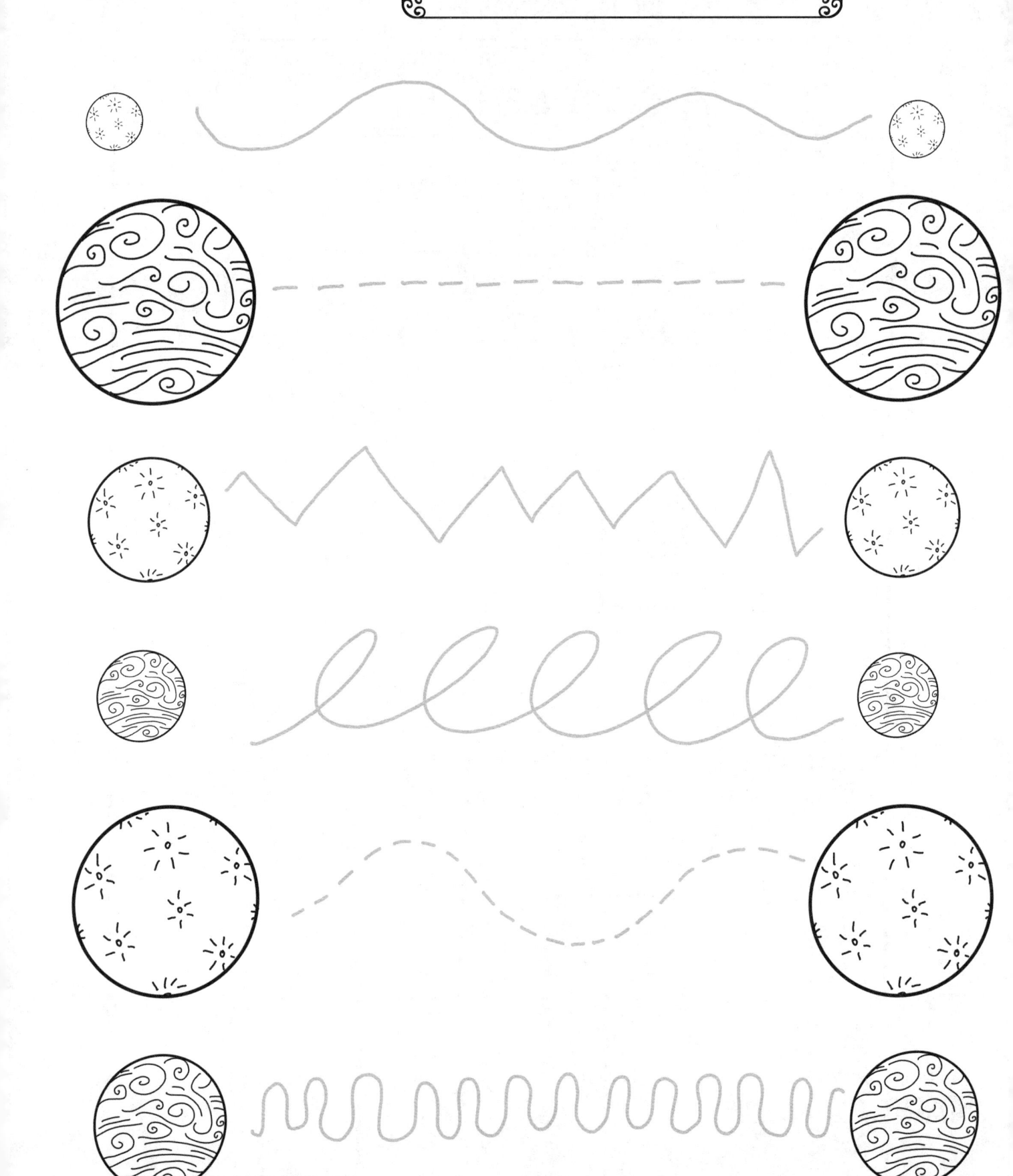

Color the number of planets listed at the beginning of each row.

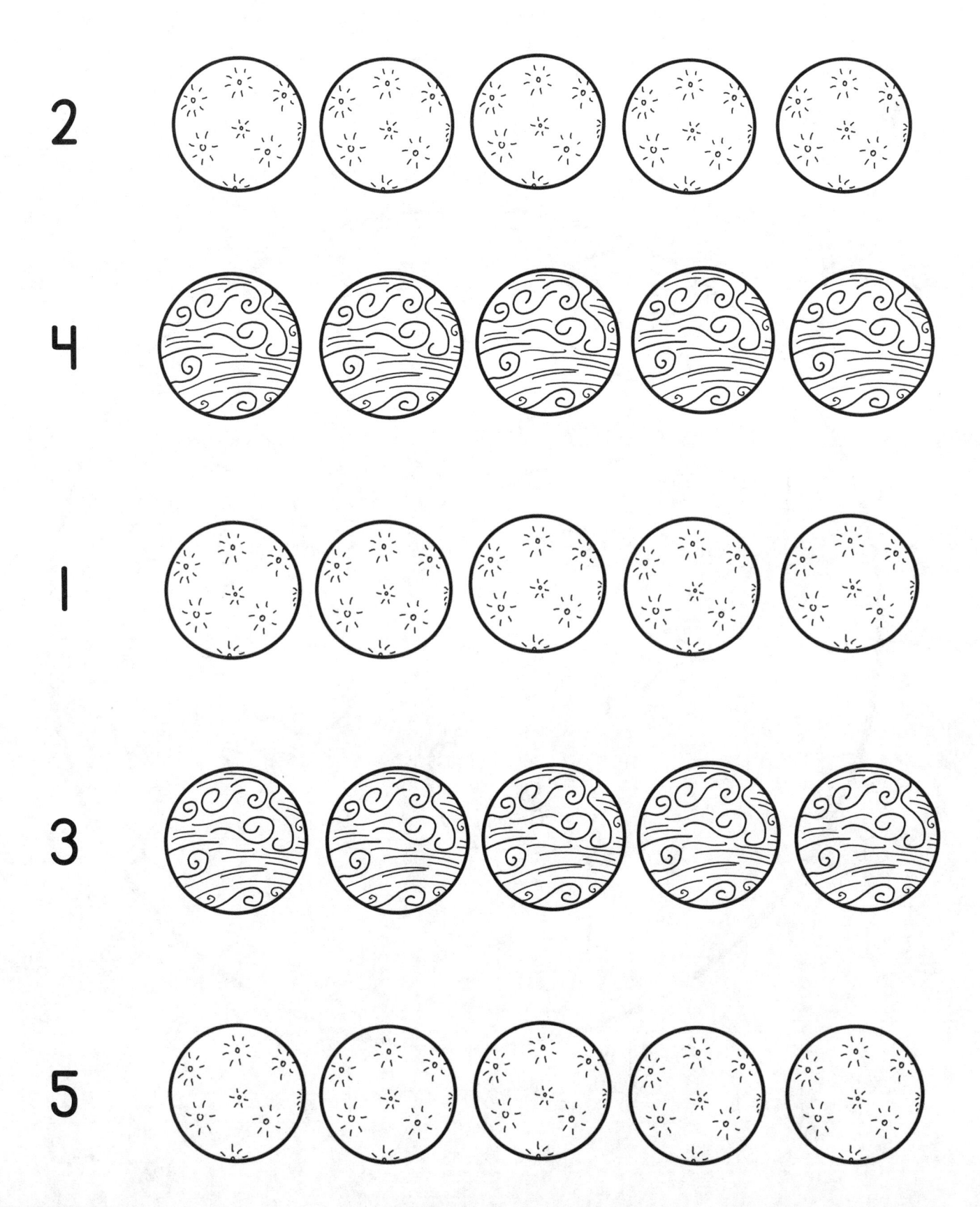

2

4

1

3

5

Lesson 6

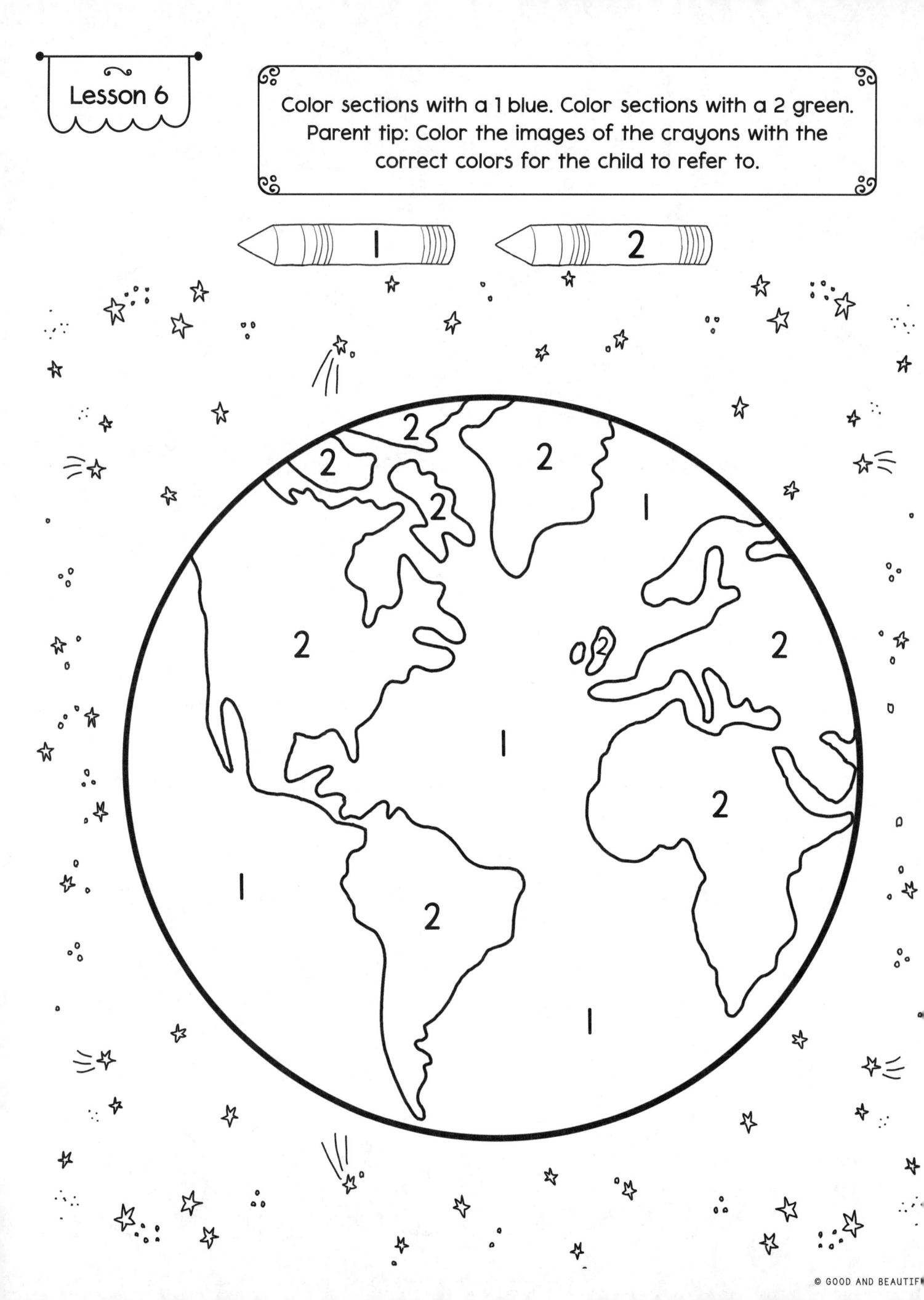

Color sections with a 1 blue. Color sections with a 2 green. Parent tip: Color the images of the crayons with the correct colors for the child to refer to.

Circle the six items in the top picture that you do not see in the bottom picture. Color the top picture.

Lesson 7

Color each phase of the moon. Then draw a line from each phase of the moon to its matching shadow.

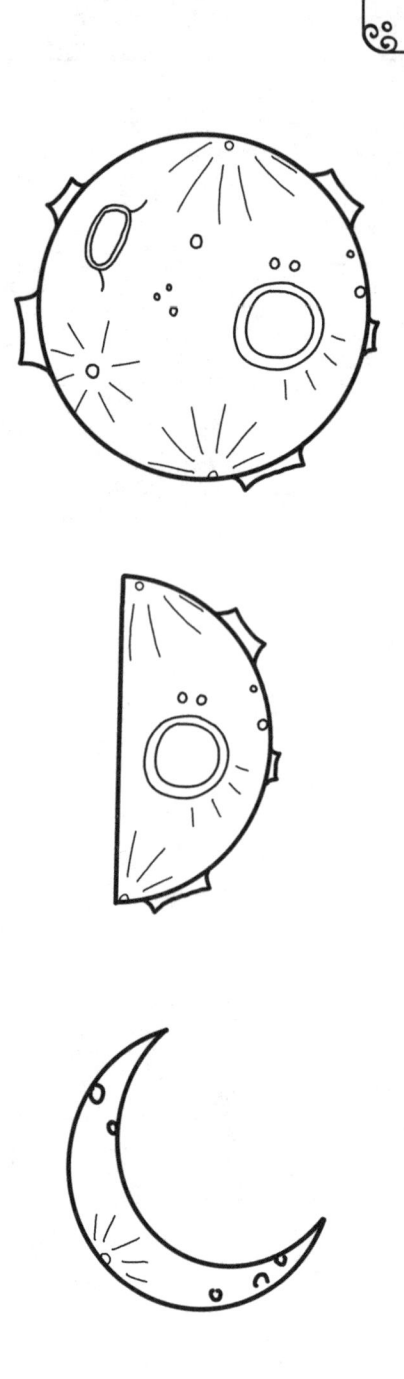

In the box below the pictures, circle the correct number to show how many of each object appear on this page. Color the pictures.

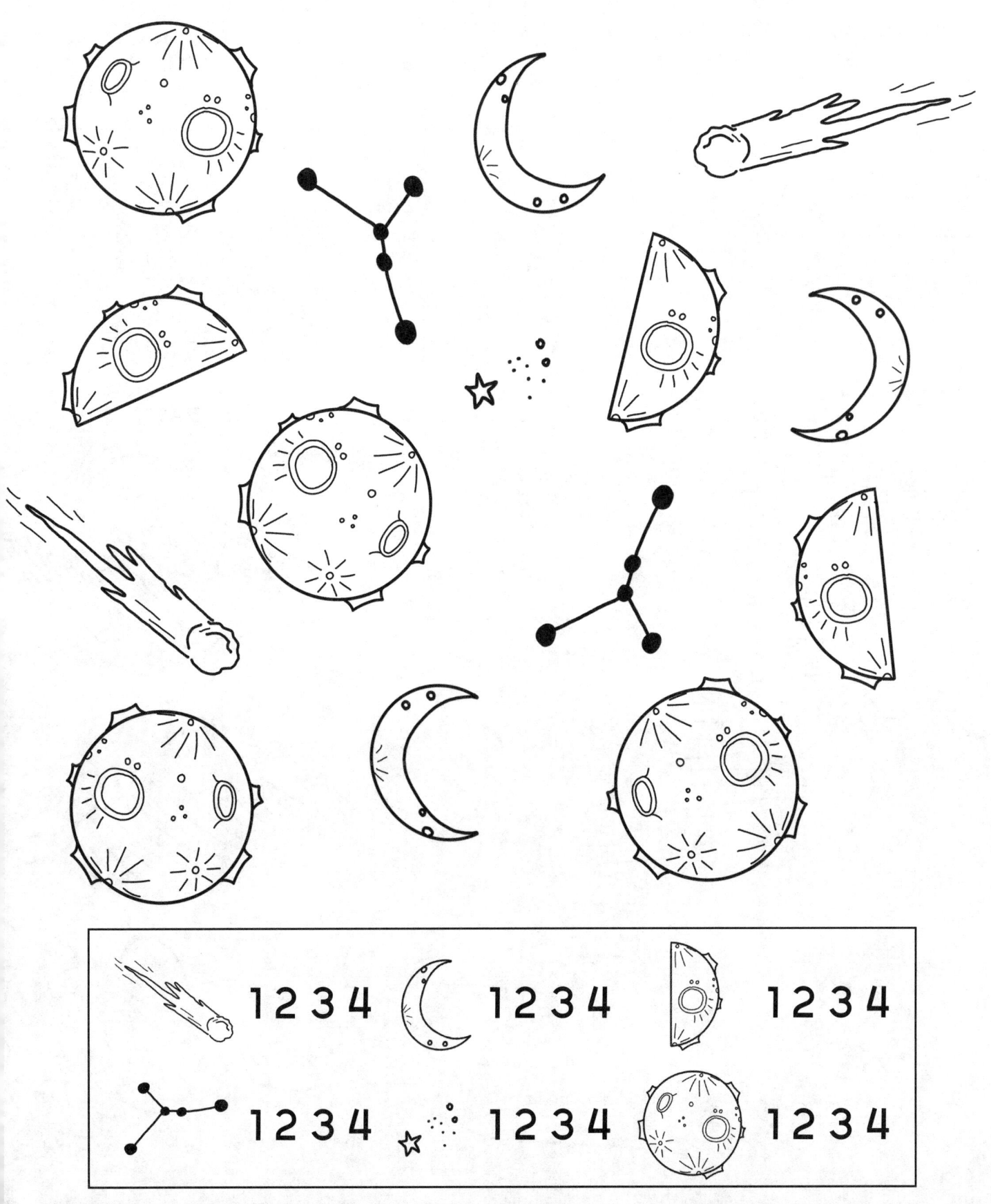

1 2 3 4   1 2 3 4   1 2 3 4

1 2 3 4   1 2 3 4   1 2 3 4

Lesson 8

Cross out the items in the key as you find them in the picture. Color the picture.

KEY

Color or circle only the different phases of the moon.

Lesson 9

Circle the picture that matches the first object in each row. Color the matching pictures.

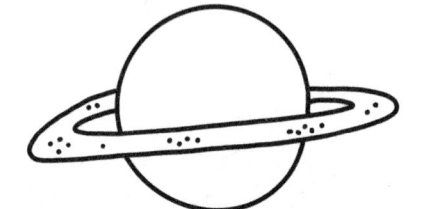

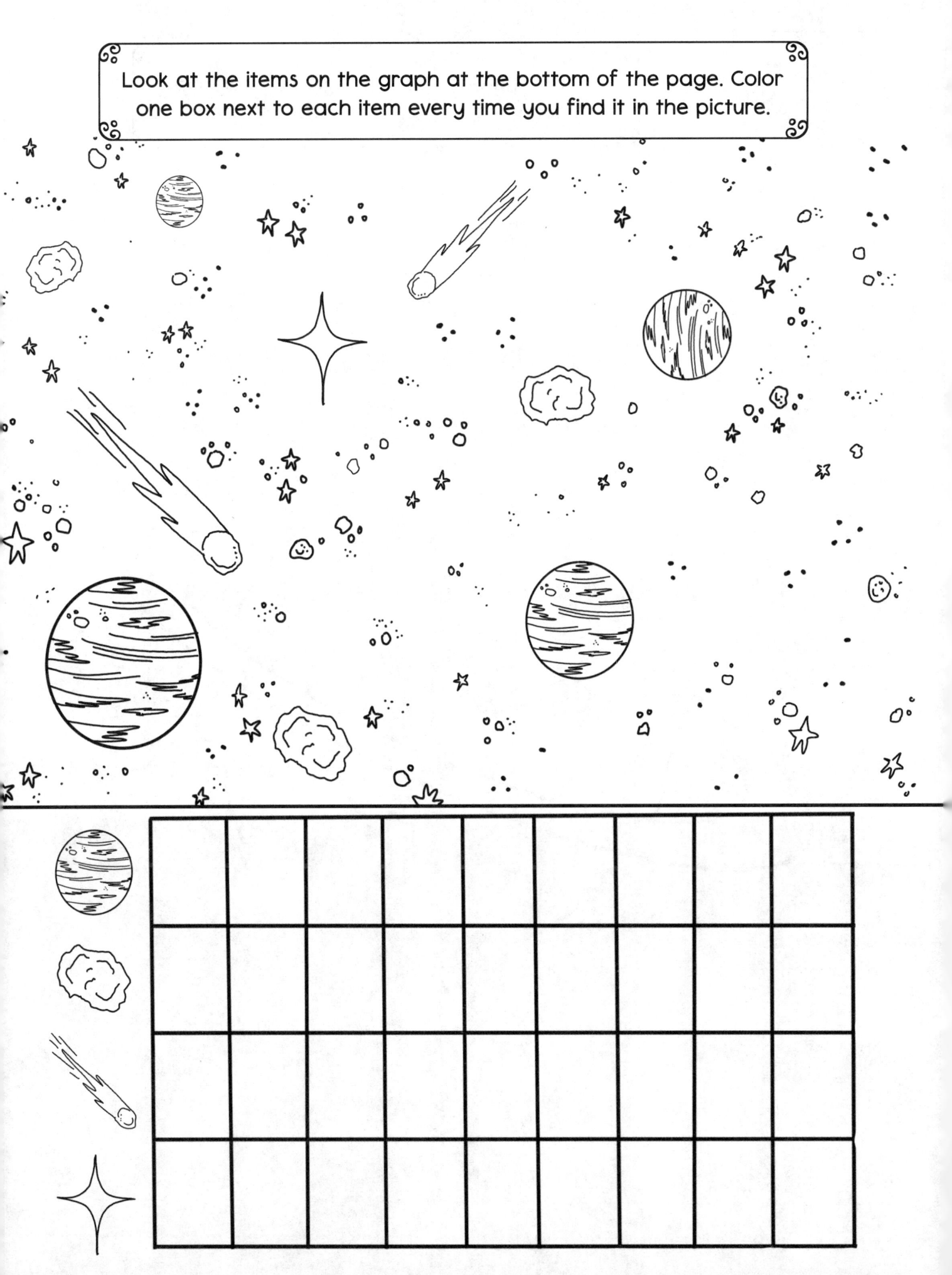

Look at the items on the graph at the bottom of the page. Color one box next to each item every time you find it in the picture.

Lesson 10

Connect the dots. Color the planet. Add stars, comets, and meteors to the picture.

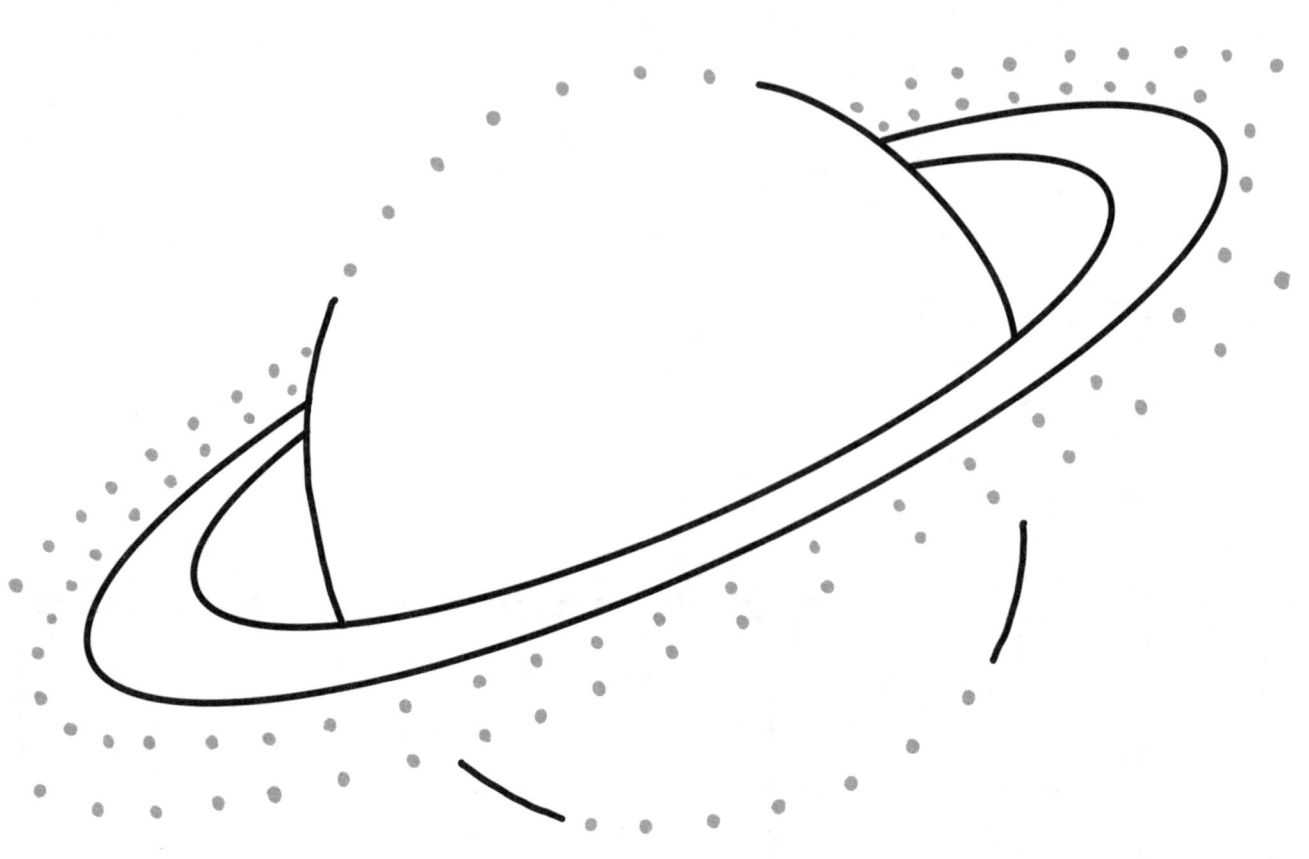

Draw a line from each planet to the number that shows how many rings are around the planet. Color the planets.

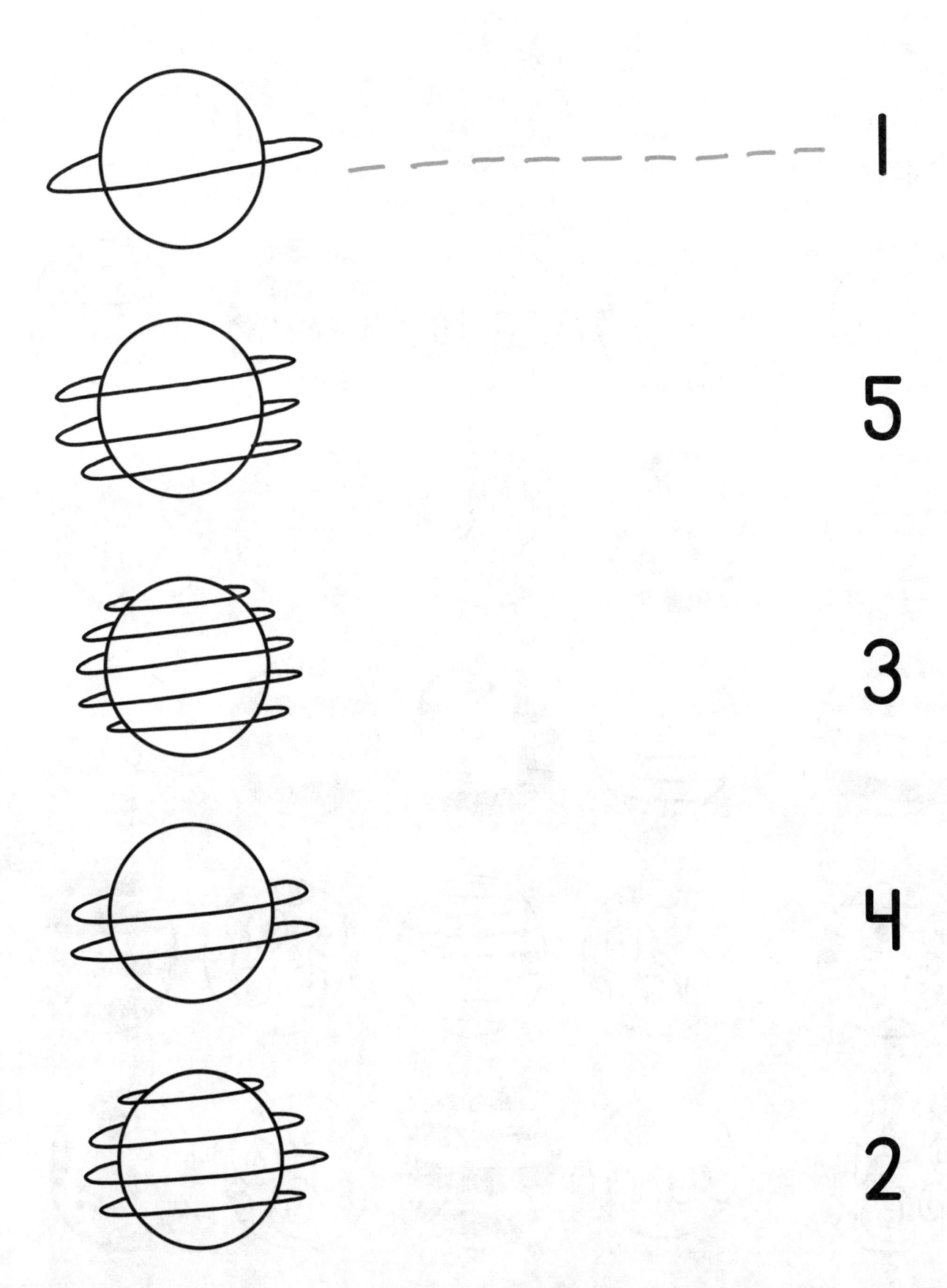

1

5

3

4

2

Lesson 11

Circle the planets that match the first planet in each row. Color the planets.

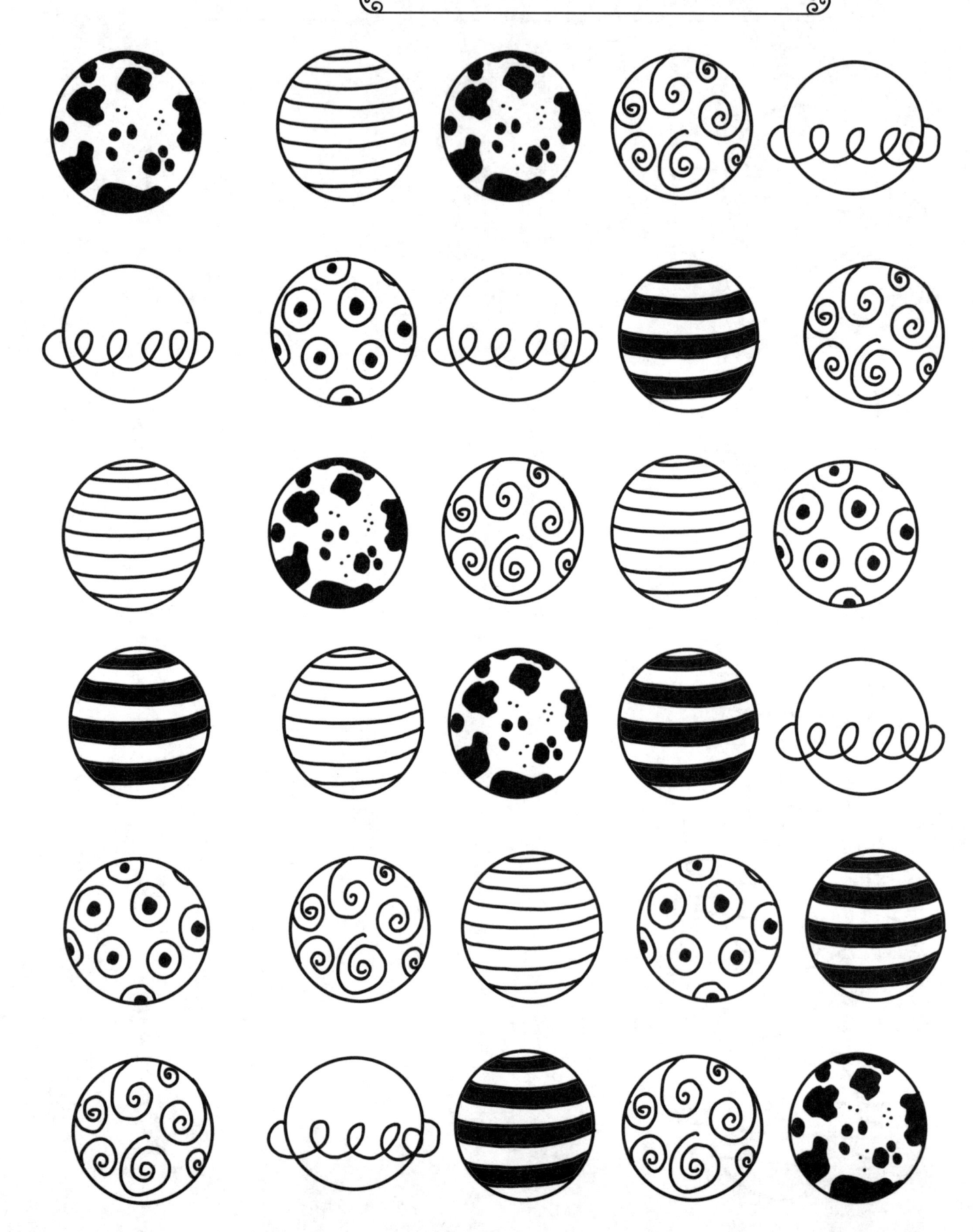

Draw rings around the planets. An example planet is below, to the left. Color the planets and stars.

Lesson 12

Find and color the parts that belong to the telescope. Color the telescope.

Circle the six items in the top picture that you do not see in the bottom picture. Color the top picture.

Lesson 13

Complete the maze to get the spaceship to the space station. Color the picture.

Connect the dots. Color the moon, stars, and comet.

Lesson 14

Color the picture.

Copy the three steps below into the box at the bottom of the page to draw a spaceship. Color and decorate your spaceship.

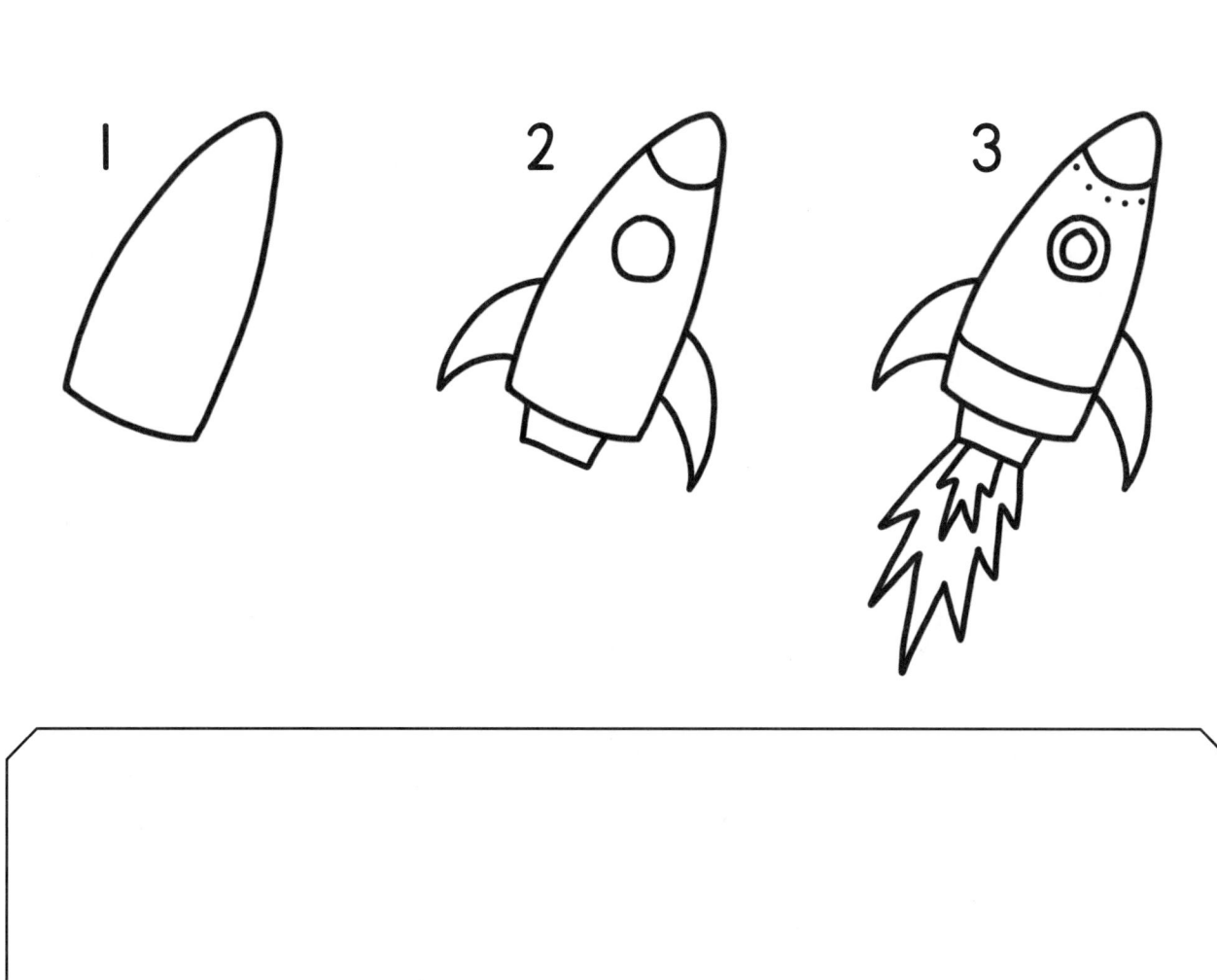

Lesson 15

Color or circle only the planets.

Draw lines to connect the matching planets. Color the planets.

Extra Doodling and Drawing Page

Extra Doodling and Drawing Page

Extra Doodling and Drawing Page